D0518535

THE
LITTLE LEMON
BOOK

Judy Ridgway

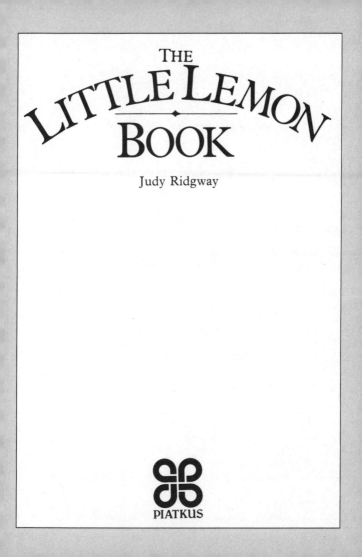

PIATKUS

Other titles in the series

The Little Green Avocado Book
The Little Garlic Book
The Little Pepper Book
The Little Apple Book
The Little Strawberry Book

© 1983 Judy Piatkus (Publishers) Limited

First published in 1983 by Judy Piatkus
(Publishers) Limited of Loughton, Essex

British Library Cataloguing in Publication Data

Ridgway, Judy
The little lemon book.
1. Cookery (Lemons)
I. Title
641.6' 4'344   TX813.L4
ISBN 0-86188-258-X

Drawings by Linda Broad
Designed by Ken Leeder
Cover photograph by John Lee

'The Lemon Tree'
© 1960 and 1961
Reedlands Music Corp and Bolder Music Inc
assigned to TRO Essex Music Ltd for World ex USA and
Canada

Typeset by V & M Graphics Ltd., Aylesbury, Bucks
Printed and bound by The Pitman Press, Bath

CONTENTS

THE LEMON TREE

When I was just a lad of ten, my father said to
    me,
'Come here and take a lesson from the lovely
    lemon tree.
'Don't put your faith in love, my boy,' my
    father said to me;
'I fear you'll find that love is like the lovely
    lemon tree.'

Lemon tree very pretty and the lemon flower is
    sweet
But the fruit of the poor lemon is impossible to
    eat.

One day beneath the lemon tree, my love and I
    did lie,
A girl so sweet that when she smiled the stars
    rose in the sky.
We passed that summer lost in love, beneath
    the lemon tree,
The music of her laughter hid my father's
    words from me.

Lemon tree very pretty and the lemon flower is
    sweet
But the fruit of the poor lemon is impossible to
    eat.

One day she left without a word, she took away
    the sun
And in the dark she'd left behind, I knew what
    she had done.
She'd left me for another, it's a common tale
    but true,
A sadder man, but wiser now, I sing these
    words to you.

Lemon tree very pretty and the lemon flower is
    sweet
But the fruit of the poor lemon is impossible to
    eat.

'THE FRUIT OF THE POOR LEMON IS IMPOSSIBLE TO EAT ...'

Of course it is not impossible to eat a lemon, but one can understand the reason behind the saying.

Lemons belong to the genus *citrus* in the family Rutacae, and all the plants in the genus contain a certain measure of citric acid, hence the term citrus fruits. The flesh of the lemon contains a particularly high proportion, which makes it extremely sour to the taste.

Lemon trees rarely grow more than 20 feet high. They are evergreen with oval-shaped leathery leaves. The young leaves have a reddish tint but this soon turns to green. Some varieties have sharp thorns in the angles between the leaves and the stem.

The lemon tree differs from other citrus fruits because it can blossom several times a year. The buds have a reddish tint, but the open petals are mainly white, and very pretty and fragrant. In some varieties the flowers grow in small clusters; in others they are much larger and occur singly.

The blossom gives rise to very highly-scented yellow fruit, oval in shape with a small nipple at one end. The rind or peel is rich in volatile oils, and underneath is a thickish layer of bitter pith. The central part of the fruit is divided into eight to ten segments which contain the flesh. The flesh consists of swollen juicy filaments, usually with a few pips or seeds.

In the language of flowers, lemon blossom stands for fidelity.

LEMONY PLANTS

Other plants share the sharp and refreshing lemony fragrance of the lemon tree but are in no way related. The leaves of Lemon Balm, Lemon Thyme and Lemon Verbena can be used in recipes which call for a lemon flavour, and are excellent in salads. The fragrances of these herbs are also used in the manufacture of soap, perfume and cosmetics, and the dried leaves of Lemon Balm and Lemon Verbena are used to make refreshing tisanes.

Lemon Oil Grass, a native of the tropics, is another plant with a lemony flavour, which in this case comes from the citronella oil contained in its leaves. The oil is used in both cosmetics and insect repellents.

'I'll be with you in the squeezing of a lemon.'
Oliver Goldsmith (1730–1774)
From *She Stoops to Conquer* Act 1 Scene 2

GROWING LEMONS
... IN COMMERCIAL ORCHARDS

L emons are cultivated in sub-tropical climates and thrive in the Mediterranean area and in California. The trees are grown in extensive orchards or groves. They are planted in regular rows about 20 feet apart, and irrigated by means of water-channels between the rows. Methods of cultivation have remained largely unchanged since the Persians first started to cultivate lemons in pre-Roman times.

Lemons are very sensitive to weather conditions and a severe frost can result in a much smaller crop than usual, but they do well on a wide range of well-drained and slightly acid soils.

Setting up a lemon grove can be quite a lengthy business, involving at least eight to ten years of tending and cultivating before a commercial crop can be expected. However, once established, a good lemon tree will yield between four hundred and six hundred lemons each year, and it will go on producing for fifty to sixty years.

Cultivated lemon trees are not grown from seed. This is because they have been developed from several different varieties and will therefore not breed true. Instead, they are propagated by taking a small bud from the desired variety and grafting it on to seedlings of other citrus species such as sweet orange or grapefruit. These seedlings are superior to the root stock of lemons as they are more uniform in size and less susceptible to disease.

The budded trees are usually about two and a half to three years old when planted in the orchard. The first fruits are formed two to three years later, and a reasonable crop may be expected in another three years.

Most orchards are harvested twice a year, though mature fruit is available on the tree in most months. The fruit is usually picked when it has grown to about 2 inches in diameter. As the best keeping and best quality fruit reaches this size when it is still green, the fruit ripens during storage and transportation.

... IN THE GARDEN AND GREENHOUSE

The climate of Northern Europe is not conducive to lemon growing, though a few enthusiasts do try their hand at it from time to time. There are, for example, lemon trees growing and bearing fruit in the public gardens of Penzance in Cornwall.

The main problem is the cold. Lemon trees need at least a summer temperature of between 60°F and 65°F and a winter temperature of not less than 55°F. However, as lemon trees are relatively small, often only around 10 or 15 feet tall, a large conservatory or greenhouse can easily accommodate young trees.

A few wealthy and enthusiastic gardeners tried to grow citrus fruits in England in the late 16th and early 17th centuries. But success was rare, particularly with the delicate lemon tree. The trees were sometimes planted in boxes on wheels so that they could be trundled into the sunshine in summer and

kept indoors by the fire in winter!

If you do decide to have a go, plant your seedlings in the early spring in loam, enriched with manure and soot or charcoal. Keep in the greenhouse and water the young plants freely throughout the summer and dose with liquid manure once a week.

... OR IN A POT

It is always possible to plant lemon pips, or seeds, and produce a perfectly manageable pot plant, but do not expect too much in the way of fruit!

Start with a reasonable-sized pot to allow room for eventual root development, and fill with moist soil from the garden or with a suitable mix of potting compost. Make sure that the soil is not too tightly packed in the pot. Take the lemon pip and push it about 1 inch down into the soil, and cover it over.

Keep the pot in the warmest part of the house and wait for the first shoots to appear. Do not allow the soil to dry out.

Continue to keep the pot in a warm place and make sure that the young seedling gets plenty of sunshine. Water from time to time.

LEMON VARIETIES

EUREKA

This variety of lemon was discovered in Los Angeles in 1858 and had its origins in lemons of Italian stock. It has been so successful that many of the lemon-growing countries of the world have taken it up and it is now grown in Cyprus, Israel and South Africa as well as in its native California. It is a juicy lemon with a medium skin and not too many pips.

LISBON

Lisbon is a widely grown variety from Israel, Cyprus and California. The fruit is very similar indeed to the Eureka but it has a slightly more pronounced nipple.

FEMINELLO

This is the most popular variety of Italian lemons and accounts for four-fifths of the crop. The fruit is oval, with a medium-thick peel and a juicy pulp.

MONACHELLO

This is another Italian variety. It is also oval in shape, but its peel is thicker than that of the Feminello lemon. However, contrary to popular belief, it is also juicier than many other lemons. The fruit is in the shops from October to March.

VERDELLO

Verdello, as its name implies, is a greener lemon than the others, and is obtained by various methods of forced growth. A Verdello is round in shape with a juicy pulp and medium-thick peel. It grows well in Italy and in the USA and is in the shops between May and September.

INTERDONATO

Ninety per cent of the Italian lemon crop comes from Sicily, and this highly disease-resistant variety was developed in Sicily in 1875 from a lemon/citron hybrid by a Colonel Interdonato. It is now grown in Cyprus, Israel and Turkey as well as in Italy. The fruit is much longer than other types of lemons and it has a very smooth thin skin.

PRIMOFIORI

A Spanish variety, pale yellow in colour, with a very thin skin and juicy flesh. The shape varies between spherical and oval, and the nipple is small and sharp. Primofiori lemons are in the shops from October to February.

VERNA

This is another Spanish variety. It has a very intensive lemon colour and a full oval shape. The flesh is juicy with very few pips. Verna is the most popular variety in Spain, and available from February to September.

FACTS AND FIGURES

World production of lemons has increased considerably over the last 25 years. In 1955, the major producers were the United States of America, Italy, Argentina, Turkey, Greece and Spain and between them they produced 908,416 tonnes; in the 1980–81 season, world production had increased to 3,198,000 tonnes. Over half of the world's lemons are grown in the Mediterranean area, the largest grower being Italy who produced 720,000 tonnes in 1981-82. In the same season, the United States produced 951,000 tonnes, of which 170,000 tonnes were exported to Canada and Europe; the total production from Argentina, Australia, Chile, Uraguay and South Africa was 410,000 tonnes.

Nearly 100 million lemons are imported into Britain every year, but most of them are used in the manufacture of goods as varied as soft drinks, concentrated lemon products, pharmaceuticals, beauty products and household cleansers. Most people buy less then two pounds of fresh lemons a year.

———————◆———————

'Know you the land where the lemon trees bloom?
In the dark foliage the gold oranges glow; a soft wind
    hovers from the sky,
The myrtle is still and the laurel stands tall – do you
    know it well?
There, there, I would go, O my beloved, with thee.'
                                        Goethe (1749-1832)

THE VALUE OF LEMONS

Lemons are exceptionally rich in Vitamin C, and they also contain Vitamins A and B and a variety of minerals. However, the main constituent of lemon flesh is water; there is a little carbohydrate and virtually no protein or fat.

Lemons contain large quantities of citric acid. A medium-sized lemon yields about 2 oz juice, containing 80-90 grains of citric acid, or about $9\frac{1}{2}\%$. Interestingly, this acid is continually being made and broken down in the body. The process is part of a vital chain of reactions called the Citric Acid or Krebs Cycle whereby energy is liberated from food. Some diets recommend foods (such as citric fruits) which interfere with this cycle and reduce the amount of food that is digested.

Citric acid helps to slow down the process of discoloration which occurs when certain foods are exposed to air. It also speeds up the separation of milk on souring, and helps to ensure a good set in jam making. The pectin content of the pith is very important in the jam making process, and is thought by many to help in the treatment of intestinal disorders. In medicine, lemon juice also acts as a tonic, it reduces fever, cures colds and coughs, and prevents scurvy. The bleaching and cleansing properties of the acid and the lemon's refreshing and distinctive smell make it useful in the house and beauty parlour.

HISTORY OF LEMONS

The original home of the lemon tree is Northern
India, but even as early as the 8th century BC
lemons had spread throughout South East Asia and
were known in China, Burma and Ceylon.

By early Roman times, and still well before the
Birth of Christ, the Jews of the Eastern Mediterranean were using citrons, first cousins of the lemon,
and including them in the ceremonies of the Feast of
Tabernacles. Citrons are still handed round at harvest
festival time for the congregation to smell and to
praise God for the sweet odours he has given to man.

Persians, or Medians as they were then known,
were by this time cultivating lemons in irrigated
orchards, and with the establishment of the Roman
Empire in the East, lemons were taken back to Rome
itself.

The Romans knew the lemon as the Median
Apple, and there are references to the fruit in both
the culinary tracts of Apicus and in contemporary
medical tracts. Lemons are also depicted in one of
the famous mosaics at Pompeii. The Romans used

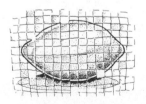

the juice to sharpen the flavour of vinegar for salads and spicy dishes, and in remedies for various ailments. It was also considered to have excellent anti-venomous properties – the writer Athenaeus states that on one occasion two men felt no effects from the bites of dangerous serpents because they had previously eaten of this fruit!

Like so many other things, lemons did not survive the decline of civilisation in Southern Europe. They ceased to be grown in the Mediterranean area and disappeared into the abyss of the Dark Ages. But, after the oblivion of that period, lemons were rediscovered, again in Persia but this time by the Arabs who propagated them around the Eastern Mediterranean and as far west as Sicily and Spain. The first Englishmen to enjoy lemons would have been the crusaders who, in 1191-92 wintered with Richard the Lion Heart in the fruit groves of Jaffa.

About 100 years later, lemons began to arrive in Britain. The first ones came from Spain in 1289, when fifteen lemons, together with seven oranges, some pomegranates and dried fruit were landed at Portsmouth as a present for Queen Eleanor, a former

princess of Castile. The rarity of such fruit made them extremely expensive and the bill for 39 lemons purchased for the Queen during her last illness in 1290 was an astonishing twenty shillings, an enormous sum in relation to the price of other food.

By 1400, the price had fallen somewhat and more frequent consignments were coming in from Spain, Portugal and Italy, on spice ships. In Elizabethan times, the price of lemons was down to a penny a piece and they were turning up as a regular culinary ingredient.

At first, lemon juice was used on its own in cooking, rather as we might use it on fish today. The juice was a favourite for sprinkling over cooked chicken or capon. Later, the juice was added to wine or seasoned vinegar to produce a new range of thin spiced sauces, while whole lemons were placed in stews with mutton or poultry.

The value of fresh lemons as an antiscorbutic, or an antidote to the skin disease scurvy, was quickly appreciated and sailors became known as limeys –

not because they used a lot of limes, but because they ate lemons to ward off the scurvy which was rife among seamen.

Neat lemon juice soon proved too sour for those who stayed on dry land. Most 16th- and 17th-century cooking catered to a population with an extremely sweet tooth. Dried fruit and sugar were added to essentially savoury meat and fish pies. Citrus fruits were soon cooked in these pies and gradually took the place of the dried fruit, but the pies were still sweetened with plenty of sugar or honey. An Elizabethan recipe for Lemon Chicken, typical of the many recorded by Gervase Markham, shows the sort of food that was enjoyed. It called for chickens to be boiled with sliced lemons; when cooked, the whole dish was sweetened heavily and then thickened with wine and egg yolks.

Surprisingly, lemons were not the Elizabethan choice as a garnish for fish. They preferred Seville oranges. However, the gourmets of the 17th century decided that lemons had a more appropriate flavour and, of course, they have remained popular ever since.

The 17th century also saw the introduction of punch from India. It became a very fashionable drink towards the end of the century when it gained strong political overtones; punch was favoured by the Whigs, while the Tories stuck to the traditional claret and sack. In the 20th century the word punch is used to denote almost any kind of hot alcoholic drink, but it originally referred only to a drink made with lemon juice, sugar, spirit, water and spices. Its name came from the Hindu word for 'five', from the five basic ingredients.

By the 18th and 19th centuries, tastes had become much more sophisticated and, on the whole, savoury and sweet dishes were kept quite separate. Lemons continued to be used to flavour some savoury dishes, but the affinity between lemons and sugar was exploited to the full in a whole range of delicious desserts. Mrs Beeton's original *Book of Household Management* lists more than forty recipes using lemons as a prime ingredient, and they include creams, jellies, blancmanges and puddings, as well as some sauces for chicken reminiscent of Elizabethan cookery, though not as sweet. Today, lemons are used in almost every kind of cookery both as a flavouring and as a garnish.

Research into vitamins and the effects of vitamin deficiencies has substantiated many of the traditional lemon remedies, and current research into the effects of Vitamin C could show that lemons are even more beneficial than we had realised.

A cold lemon drink popular in the 18th century was shrub. This was Eastern in origin, being a favourite Arab drink, and was made from white wine, brandy or rum, lemon juice, lemon peel and sugar.

CHOOSING LEMONS

It is difficult to tell simply by looking at a lemon exactly how thin or thick the peel is going to be and how many pips the fruit will contain. Nor are thick skins necessarily to be avoided, as some thick-skinned varieties such as the Italian Monachello are very juicy indeed.

With most varieties, the weight of the lemon is a good indication of its juiciness. The skin and pith weigh a good deal less than the flesh and juice, and so if a lemon feels heavy for its size it will probably be juicy. A look at the skin will also tell you something about the fruit; rough surfaces with large pores often indicate thick skins.

Go for:
- ★ Brightly coloured, fresh skinned fruit
- ★ Fruit with a fine skin and small pores
- ★ Fruit which is heavy for its size
- ★ Fruit which feels fairly soft to the grip

But remember the exceptions:
- ★ Green Verdello lemons from Italy and the USA
- ★ Blackened and pimply lemons from the West Indies (West Indian lemons are not very common, and they certainly do not look very nice, but they are almost completely seedless and full of juice)

Price can also be a helpful guide; it is best to avoid fruit that appears to be ridiculously cheap – the seller probably knows what is wrong with them even if you don't! Do not worry about the size of the lemons. Size has no direct bearing on quality, and a small lemon will be cheaper than a large one.

Blemishes or marks on the skin are usually of no importance as they are caused by the weather and do not reflect the condition of the fruit. Some lemons appear to be particularly bright and shiny, due to a coating of a protective substance which, although perfectly harmless, may contaminate the flavour of anything in which the peel is used. Californian lemons are nearly always so treated, so check where your lemons have come from and always wash them well.

STORING FRESH LEMONS
... THE MODERN WAY

L emons can be kept for two or three weeks in the right conditions. Place them in a cool and well-ventilated place such as a larder or cellar, making sure that there is plenty of space for the air to circulate between the fruit.

*   Never put a lemon in the fridge; chilling can reduce the amount of juice a lemon will yield by up to half.
*   Never leave any damaged fruit with the good ones; the bad fruit will soon affect the whole batch.

... THE OLD FASHIONED WAY

Our grandmothers thought they knew the answers to the problem of longer-term storage of fresh lemons. An old household encyclopaedia suggests the following methods, but no indication is given of the increase in shelf life to be expected.

*   Smear the lemon all over with the white or yolk of an egg and place it on a shelf to dry.
*   Keep the lemon in a jar of water and renew the water every day.
*   Place the lemon in a box, cover it with clean dry salt, and store it in a cool dry place.

Preparing Lemons

Wash lemons well in hot water immediately before using to remove the dust and dirt as well as any artificial polishing or preserving agent. The immersion in hot water also helps the juice to run more freely if you are planning to squeeze the lemon.

If a lemon feels particularly hard, wrap it in a warm dish cloth or place it in a warm oven for a couple of minutes, then squeeze.

Squeezing

By far the simplest way to squeeze the juice from a lemon us to cut it in half and use a mechanical or electrical squeezer. However, if you do not have one of these gadgets, soften the fruit by squeezing it between your hands or rolling it back and forth over a hard surface, then cut a small hole in the top and squeeze out the juice. Or cut the fruit in half and squeeze both halves separately.

Peeling

Use a very sharp knife or a potato peeler to peel off the zest of a lemon, leaving as much of the pith behind as possible. Special Julienne knives can be purchased for peeling off long thin strips of the outer peel. The zest can also be grated off the pith with a very fine-gauge grater; a coarse one will take off too much pith with the zest.

SKINNING

Some recipes call for sliced, skinned lemons, and the whole of the outer peel and pith can be quite difficult to remove. To loosen the skin, place the fruit in a pan of boiling water for a minute or two, then remove and allow it to cool slightly. Next, make four vertical cuts in the skin from top to bottom and peel off each quarter. The skin should come off quite easily.

USING LEMONS

There is no wastage in a lemon; every little bit can be pressed into service. So next time you squeeze one, remember to keep the skins and use the zest for flavouring and the rest to clean your vegetable-stained hands.

THE WHOLE LEMON
(sliced or chopped)

* For making lemonade, squash and syrups
* For making marmalade and lemon jelly
* For making pickles and chutneys
* For garnishing meat and fish dishes
* For flavouring a large casserole of chicken or lamb (the Elizabethan method)

THE JUICE
(squeezed and strained)

* For flavouring soups, stews, vegetables and fish dishes, sauces, puddings and biscuits
* For seasoning salads and mayonnaise, in place of vinegar
* For making lemon curd
* For adding extra acid to jam and other preserves
* For making long drinks and cocktails
* For making home remedies for coughs and colds
* For bleaching and cleansing hair and skin
* For making household bleach and cleansers

Do not use the juice in baking unless the recipe demands it; it makes bread and cakes rather heavy.

THE OUTSIDE PEEL OR ZEST
(cut into thin strips)

* For making flavourings and garnishes for drinks and cocktails
* For making Julienne strips to garnish sweet and savoury dishes
* For drying for use as a store cupboard flavouring

(grated)

* For flavouring bread, cakes and biscuits
* For flavouring savoury dishes and desserts

THE SKINS
(whole and sliced)

* For making candied peel
* For holding sorbets and other dishes (in a frozen half shell)
* For cleaning vegetable-stained hands

THE PITH AND PIPS

* For adding extra pectin to jams and marmalade

FREEZING LEMONS

PEELED LEMON SLICES

To freeze: Peel the lemons, taking care to remove all the pith. Cut into thin slices and remove the pips. Pack the slices into rigid containers and cover with a light syrup made up of 4 oz sugar to 1 pint water. Fast-freeze until hard and store for up to a year.

To use: Thaw at room temperature and use as a garnish or flavouring.

UNPEELED LEMON SLICES

To freeze: Wash the lemons well and cut into thin slices. Place on a tray and open freeze in the fast-freeze section of the freezer. As soon as they are hard, pack into polythene bags and keep in the storage area for up to a year.

To use: Use from frozen as a garnish.

GRATED LEMON PEEL

To freeze: Place the grated peel in very small containers or bags and freeze it as it is. Alternatively, freeze teaspoonfuls of grated peel with a little water in ice cube trays. When frozen, remove from the trays and pack the cubes into polythene bags.

To use: Use from frozen, adding the cubes to dishes such as soups and casseroles and the dry frozen peel to other dishes.

LEMON JUICE

To freeze: Pour into ice cube trays and freeze. When frozen, remove from the trays and pack into polythene bags.

To use: Use from frozen in liquid dishes and drinks.

PRESERVING LEMONS
DRIED LEMON PEEL

This is a very simple way of preserving the flavour of lemons.

Peel the lemons and remove as much of the pith as possible. Wash the peel, dry it and place on a baking tray. Bake very slowly in a low oven until the peel is crisp all over. Leave to cool and pack in an airtight jar. Alternatively, break into small pieces and roll with a rolling pin or grind to a fine powder in an electric blender. Store in an airtight jar in a cool, dry and dark place.

Use pieces of dried peel to flavour casseroles, soups and stews. Use the powder to flavour any kind of cooked dish, but use sparingly as it can be pretty strong.

A piece of dried peel placed in a small storage jar of sugar will produce a lovely lemon-flavoured sugar for use in custards, creams and syllabubs.

26

CANDIED PEEL

This is another delicious way to preserve lemon peel. Candied peel can be chopped and used to flavour bread, cakes and biscuits, or it can be dipped in melted chocolate and served with petit fours. It is rather soft, and quite delicious.

2 thin-skinned lemons
6 oz sugar
water

Peel the skins off the lemons and use the flesh in a lemon salad. Cut out any blemishes from the skin, and cut into $\frac{1}{3}$ in x 3 in lengths. Place the lemon peel in a pan, cover with water and bring to the boil. Simmer for 15 minutes and drain. Cover the lemon peel with fresh water and return to the boil. Simmer for a further 30-45 minutes until tender. Drain and keep on one side.

Place 5 oz sugar in a thick-based pan with 2 tablespoons water. Bring to the boil, stirring all the time, and heat to 238°F/115°C. Add the cooked peel and simmer for a further 20-30 minutes until all the syrup has been absorbed and the mixture is full of crystals and looks thick. Stir frequently during this phase to stop the mixture sticking to the base of the pan and burning.

Separate out the strips and toss in the remaining sugar. Place on a wire rack to dry. Do not roll in sugar if planning to dip in melted chocolate.

Makes about 30-35 strips

SAVOURY DISHES
EGG AND LEMON SOUP

This is a popular Greek soup, where it is known as Avgolemono Soup. If possible, make your own chicken stock by boiling up a boiling fowl or even the carcass of a roast chicken. Reduce the resulting stock to give a really good flavour.

2 pints well-flavoured chicken stock
2 oz risotto or medium-grain rice
juice of 2 small lemons, strained
2 large eggs, beaten
salt and pepper

Place the stock in a pan with the rice, bring to the boil and simmer for 15 minutes until the rice is soft.

Place the lemon juice and eggs in a soup tureen and beat well with a wire whisk. When the mixture is light and frothy, carefully pour in the hot soup, whisking all the time. Season to taste and serve at once.

Serves 6

BUTTER BEANS WITH LEMON AND CREAM SAUCE

Most of the major lemon growing areas have their own special dishes featuring lemons. This recipe comes from South America. If you want to use dried butter beans, soak them overnight in cold water, drain and cover with fresh water. Bring to the boil and simmer for about an hour until tender.

2 x 7 oz cans of butter beans
juice of 1 lemon
4 tablespoons double cream
grated lemon rind from 1 lemon
2 tablespoons chopped parsley
salt and pepper

Pour the contents of the cans of beans into a pan with the lemon juice. Bring to the boil and simmer gently for 5 minutes. Drain, retaining 4 tablespoons of the cooking liquor in the pan. Place the beans in a serving dish and keep warm.

Add the cream to the pan with the cooking liquor from the beans. Stir in the grated rind and parsley and bring to the boil. Simmer for 3 minutes until the sauce is reduced by about half. Season to taste and pour over the beans. Serve at once.

Serves 4

MEDALLIONS OF VEAL WITH LEMON

1½ lemons
1 teaspoon sugar
water
1 lb fillet of veal cut into 4 medallions
salt and pepper
1 oz butter
4 fl oz dry white wine
2 tablespoons chopped parsley

Pare the rind off the whole lemon and cut into thin strips. Place the rind in a pan and cover with water. Bring to the boil and cook for 5 minutes. Drain and refresh the peel under cold water. Drain and return to the pan. Add the sugar and 3 tablespoons water. Bring to the boil and simmer until all the water has evaporated.

Remove the pith from the pared lemon and separate the fruit into segments. Remove the skins.

Trim any excess fat off the veal and season well. Melt the butter in a frying pan and fry the veal for about 4 minutes on each side. Remove from the pan and keep warm. Pour in the wine and the juice from the remaining ½ lemon and stir well over a moderate heat. Add the parsley and boil until reduced to generous tablespoons.

To serve, arrange the veal on a plate and decorate the top of each medallion with the segments of lemon and a little of the rind. Pour the sauce over the top.

Serves 4

LEMON BUTTER

This quick and easy-to-make flavouring can be used to pep up any kind of poached or grilled fish. It also makes an unusual accompaniment to steaks, hamburgers or chops.

4 oz butter, softened
grated rind of 2 lemons
2 tablespoons freshly chopped parsley

Blend the butter with the lemon rind and parsley. Shape the mixture into a sausage and roll up in damp greaseproof paper, or spread it out, to a thickness of about $\frac{1}{4}$ in between two sheets of damp greaseproof paper. Place the flavoured butter in the fridge to chill.

To use, cut the sausage shape into slices, or use fancy cutters to cut out shapes from the flat slab. Place on the food just before serving.

LEMON AND HERB BREAD

Make the lemon butter as above and add a pinch of mixed herbs with the parsley. Slice a loaf of French bread almost all the whole way through and spread the butter on the slices. Wrap the loaf in foil and bake in a very hot oven for 5 minutes. Serve at once with soup, casseroles or pasta.

Lemon Desserts
Lemon Cream Royale

Just as Charles II was partial to a dish of syllabub, so William IV liked his bowl of Lemon Cream. In those days, flower waters were much more common than they are today, and most self-respecting house-keepers used to infuse the orange blossom and rose petals from their gardens. Nowadays you can buy flower waters in some chemists, herbalists or Eastern grocery shops.

juice of 2 small lemons (2½ fl oz)
3 oz caster sugar
1 teaspoon orange flower water or rosewater (optional)
water
3 egg whites

Place the lemon juice and sugar in a pan with the flower water if used, and 5 or 6 teaspoons plain water. Heat gently until the sugar dissolves. Leave to cool.

Whisk the egg whites until they are stiff and stir into the lemon mixture. Cook in the top of a double pan over a low heat for 5-6 minutes until the mixture is really thick. Stir all the time with a wooden spoon.

Serve hot or cold with thin wafers or crispy biscuits.

Serves 4

LEMON MOUSSE

This elaborate cream or mousse is made with whipped cream, and to save cooking it is set with gelatine. If time is really short, use thick quark (skimmed milk soft cheese), sieved cottage cheese or cream cheese in place of cream and omit the gelatine. The mousse will set in about 2-3 hours with gelatine, and about 1 hour without.

½ pint lightly whipped cream
¼ pint lemon juice (juice of 3 large lemons)
grated rind of 1 lemon
2 oz caster sugar
½ oz packet of gelatine
2 tablespoons water
2 egg whites, stiffly beaten

Whisk the cream, lemon juice and rind together and stir in the sugar. Mix the gelatine and water in a cup and place in a pan of boiling water. Stir until the gelatine dissolves. Leave to cool for a little and then add to the cream and lemon mixture. Fold in the egg whites and spoon the mixture into a large serving bowl or into individual glass dishes. Place in the fridge to set.

Serves 6

LEMON MERINGUE PIE

One of the most popular of all puddings.

6 oz shortcrust pastry
grated rind of 2 lemons
6 oz granulated sugar
½ pint water
juice of 2 lemons
3 tablespoons cornflour
3 eggs, separated
1 oz butter, cut in small pieces
4 oz caster sugar

Roll out the pastry and line an 8-9 inch flan case. Bake blind at 400°F/200°C/Gas 6 for about 15 minutes until crisp and golden. Cool and remove from case.

Place the rind in a pan with the sugar and water. Dissolve the sugar over a low heat, then bring to the boil. Mix the lemon juice with the cornflour, pour on the syrup, stirring. Beat in the egg yolks, one at a time, together with the butter. The mixture should be thick enough to coat the back of a spoon. If not, return to the pan and cook without boiling.

Whisk the egg whites until stiff, add 2 oz caster sugar and continue whisking until stiff.

Spoon the lemon mixture into the pastry case and top with whipped egg white. Sprinkle with remaining sugar, and bake at 300°F/160°C/Gas 2 for 20-30 minutes until the meringue is golden and crisp. Serve warm.

Serves 4–6

BAKING WITH LEMONS

LEMON SODA BREAD

This recipe makes a tangy bread which should be eaten fairly soon after baking.

8 oz plain flour
1½ teaspoons baking powder
½ teaspoon bicarbonate of soda
½ teaspoon salt
2 oz butter
1 oz sugar
1 oz wheat germ
grated rind of 1 lemon
1 egg, beaten
juice of 1 lemon
1-2 tablespoons water

Sift the flour, baking powder, bicarbonate of soda and salt into a mixing bowl. Cut the butter into small pieces and rub into the dry ingredients. Stir in the sugar, wheat germ and lemon rind. Beat in the egg, lemon juice and sufficient water to make a scone-like dough. Shape into a round loaf and score the top. Place on a baking sheet.

Bake at 350°F/180°C/Gas 4 for 1 hour until cooked through. If a skewer comes out clean the bread is cooked.

Makes 1 loaf

LEMON PICKLES AND PRESERVES
INDIAN LEMON PICKLE

Indian cooks have long known that sharp-flavoured lemons make an excellent pickle. This recipe calls for hot sunshine to finish it off, but a warm airing cupboard will probably work almost as well.

2 lb or 6 large lemons, washed
4 tablespoons salt
1 teaspoon turmeric
2 teaspoons garam masala
1 teaspoon chilli powder
a few green chillis (optional)

Cut the lemons into small chunks and remove all the pips. Mix in the remaining ingredients and place in screwtop jars. Keep in the sun for a week, giving the jars a good shake every day. The pickle will be ready to eat when the skins of the lemons are tender.

Store in a cool dry place and shake from time to time. It will keep for 2 to 3 months.

Makes 4 x 1 lb jars

36

LEMON CHUTNEY

This is a delicious condiment with an Indian flavour. In the 19th century, army families returning home from India brought back with them a variety of local recipes, many of which have been preserved in Victorian notebooks. Originally eaten with curry, chutneys like this one are also good with cold meats, hamburgers or fried chicken.

2 lb or 6 large lemons
2 lb onions, chopped
1 lb cooking apples, cored and chopped
8 oz raisins
1 pint malt vinegar
1½ teaspoons cinnamon
½ teaspoon allspice
pinch nutmeg
½ teaspoon coriander
salt and pepper
1 lb brown sugar

Wash the lemons and cut into small chunks. Remove the pips and place in a pan with all the remaining ingredients except the sugar. Bring to the boil, reduce the heat and simmer for 2 hours. Remove the lid, add the sugar and boil for a further 30-45 minutes until the mixture is soft and pulpy.

Spoon into jars with vinegar-proof screwtop lids. Screw down the lids and leave to cool. Store in a cool dry place for up to 6 months.

Makes approximately 5 lb

LEMON MARMALADE

As early as the 15th century, cooks knew how to preserve citrus fruits in sugar, and consignments of an early version of marmalade, called succade, were shipped into Britain from Spain and Portugal along with the fresh fruits. This, however, is an up-to-date recipe.

2 lb or 6 large lemons
4½ pints water
3 lb sugar

Scrub the lemons, peel them thinly and cut the rind into strips. Remove the pith from the lemons and keep on one side. Cut the flesh into pieces, removing the pips and any tough membranes.

Put the pith and pips into a small muslin bag and place in a preserving pan with the chopped flesh and bits of rind. Add the water and bring to the boil. Simmer for about 1½-2 hours until the rind is soft and the liquid has reduced by between a third and a half of its original volume.

Remove the bag of pith and pips and stir in the sugar. When the sugar has dissolved, bring the mixture to a fast boil. Continue boiling until a sugar thermometer registers 220°F/104°C. Remove any scum and leave it to stand for 1-2 minutes away from the heat. As soon as a thin skin begins to form, pour the marmalade into warmed, clean dry jars. Cover at once with waxed discs and leave to cool. Cover and store in a cool dry place for up to 2 years.

Makes 5 x 1 lb jars

LEMON CURD

Lemon Curd, or Lemon Cheese as it is sometimes known, is not really a preserve for it will only keep for a few weeks in a cool dry place, or up to 2 months in the fridge.

6-8 lemons
8 oz butter
1½ lb sugar
5-6 eggs

Grate the rind from 6–8 lemons, then squeeze the lemons to make ½ pint juice and strain. Add the rind to the juice in the top of a double pan. Add the butter and sugar and place over gently simmering water. Stir until the sugar has dissolved and the butter melted. Beat the eggs and strain into the mixture. Continue cooking, stirring all the time, until the mixture is thick enough to coat the back of a wooden spoon. Do not allow the mixture to boil or it will curdle.

When the mixture has thickened, pour into clean, dry and warm jars. Cover with waxed discs and leave to cool. Cover and store in a cool place.

Makes approximately 3 lb

LEMON DRINKS
LEMONADE

Lemonade was a French invention of the 17th century and the basic ingredients are the same today as they were then.

3 lemons
2 oz sugar
1 tablespoon honey
2 pints boiling water

Pare the rind off one of the lemons and place in a heatproof jug. Squeeze all the lemons and add the juice to the rind. Stir in the sugar, honey and boiling water and continue stirring until the sugar and honey have dissolved. Cool, strain and chill.

CITRON PRESSÉ

A quick and simple method of serving lemon juice has been adopted in those countries which grow the fruit. The lemons are squeezed and juice poured into a jug, which is then topped up with iced water, and sugar added to taste.

'The Persian's Heaven is easily made,
T'is but black eyes and Lemonade!'
Sir Thomas Moore (1478-1535)

LEMON GARNISHES

Garnish with lemon to add both flavour and colour to food and drinks.

WEDGES

To prepare: Cut the lemon in half lengthways, and then cut again into quarters or sixths. Cut off any thick segmentation or pith in the centre.

Use: To decorate all kinds of savoury dishes. (Special small wedge-shaped lemon squeezers can be bought and used to squeeze the juice over the food.)

SLICES

To prepare: Cut the lemon into slices and remove all the pips. For a more attractive effect, cut notches out of the peel.

Use: To decorate all kinds of savoury dishes.

SPIRALS

To prepare: Using a potato peeler, pare off strips of peel in a continuous spiral. Make sure the pieces are free of pith.

Use: To decorate long drinks and the plain surfaces of desserts.

TWISTS

To prepare: Cut the lemon into thin slices. Make a radial cut into the centre. Twist the slice to give a spiral effect.

Use: To decorate all kinds of savoury dishes and place on the rims of glasses or on cocktail sticks.

BUTTERFLIES

To prepare: Cut the lemon into very thin slices. Make 4 radial cuts almost to the centre. Remove two of the resulting triangles to leave a butterfly shape.

Use: To decorate savoury and sweet dishes.

JULIENNE STRIPS

To prepare: Using a Julienne knife, cut long thin strips of peel from the lemon. Use as they are or boil in water or in water and sugar to soften and sweeten.

Use: To decorate savoury and sweet dishes.

OTHER USES FOR LEMONS

KEEPING THE COLOURS BRIGHT

Lemons are extremely useful in preventing the discoloration of other foods, many of which brown very quickly once the flesh is cut open and exposed to the air, or lose their colour in cooking. These changes in colour are caused by enzymes, and the ascorbic acid in lemon juice helps to slow down their action.

★ As soon as they have been cut and sliced, drop apples, carrots or bananas into a bowl of water which contains a little lemon juice.

★ Rub the cut surface of halved avocados with neat lemon juice, or toss the sliced fruit in a lemon and oil dressing.

★ Add lemon juice to avocado or banana purée when making soups, starters and desserts. One to two teaspoons of juice per fruit should be sufficient to prevent discoloration for a couple of hours.

★ Add lemon juice to the cooking liquor to keep salsify really white. Store cooked salsify in the fridge in its cooking liquor.

★ Cook delicately-coloured vegetables like artichokes, asparagus and courgettes with a little lemon juice in the water to help them keep their natural fresh green colour. Take care not to cook the vegetables for too long.

★ Add lemon juice to the cooking liquor of button mushrooms, or sprinkle lemon juice over raw mushroom salad; the mushrooms will remain much whiter.

★ Marinate raw fish, such as fine-grain haddock, plaice or lemon sole, in lemon juice for 4 to 6 hours, keeping it in the fridge. The fish will turn really white and will be tender enough to eat in an hors d'oeuvres. The action of the lemon juice tenderises the fish.

★ Add lemon juice to royal icing to achieve a really white look.

SOURING CREAM AND MILK

If a recipe calls for soured cream and you do not have any to hand, simply sour some fresh cream with a little lemon juice. Allow 1 tablespoon fresh lemon juice to each ¼ pint of cream. Mix together and leave to stand for five minutes. Fresh milk can be soured in the same way.

SETTING JAMS

Some fruits have a low acid content and this can impare the set of the jam. Keep some lemons handy during the late summer and early autumn jam-making season and you will not have any difficulties – simply add 2 tablespoons lemon juice to each 4 lb of fruit. Jams such as apple, apricot, cherry, logan-berry, marrow, peach, pear, raspberry, rowanberry and strawberry will all benefit from the addition of lemon juice, and so will sweet apple, raspberry and strawberry jellies.

Many fruits are low in pectin and this deficiency makes fruit jams and jellies particularly difficult to set. The pith and pips of lemons are packed with pectin, and put in a muslin bag and suspended in the fruit mixture during the cooking period, they will greatly help the set. Remove the bag before adding the sugar. Try this method with sweet apple jelly, marrow jam, pear jam and strawberry jam and jelly.

HERB JELLIES

Everyone knows about mint jelly, but not so many people know that jellies can be made with other herbs. Parsley jelly was popular in Victorian times, and tarragon, sage and basil can be used to make jellies with their own distinctive flavours. Herb jellies are easy to make but they do need lemon juice to help the set. Use cooking apples to make the basic juice and then continue in the normal way.

LEMONS IN HEALTH AND MEDICINE.

L emons are exceptionally rich in Vitamin C, which is recognised as being an important factor in the prevention of infection. Though experts continue to disagree as to the exact role the vitamin plays in the prevention and cure of colds and flu, our ancestors had no such doubts about the efficacy of lemons and used them in all manner of remedies, from treating scurvy to curing headaches, hangovers and even loss of voice. The juice was considered to be particularly good for relieving thirst as well as for cooling fever; it was thought to aid the digestion, probably because of the ability of citric acid to help in the break-down of fatty foods.

Modern pharmaceutical manufacturers seem to be convinced about the value of lemons; they are included in both liquid and powder remedies for coughs, colds and flu. Lemons are used in lozenges and liquids to mask the taste of other active ingredients, and in liquid medicines the citric acid content can act as a preservative. Citric acid is used to help control the acid/alkaline balance of some medicines; it is also found in acid eyewashes for treatment after accidental alkaline splashes in laboratories and factories.

Syrups and cordials containing lemon juice used to be specially brewed to combat coughs and sore throats. Many people today find these old remedies most effective.

LEMON COUGH MIXTURE

Ingredients: juice of 1 large lemon, strained
1 tablespoon glycerine
4 tablespoons honey

To mix: Place a cup containing all the ingredients in a pan of hot water. Stir until the honey dissolves and the mixture is heated through.

To use: To treat coughs and sore throats, take two teaspoonfuls before going to bed, and as required during the day. Keep the mixture covered and reheat for each dose.

SICILIAN REMEDIES

A sophisticated remedy for flu and colds is to drink a steaming hot glassful of whisky, lemon and honey, or, as one Sicilian exporter of lemons recommends, brandy, lemon and honey. (The French version is to mix lemon juice with hot water and sugar.)

The Sicilian cure for 'Mediterranean Tummy' is to drink (if you can!) the juice of 4 lemons, including the oil squeezed from the peel.

LEMON BARLEY WATER

In some cases of kidney or bladder infections, the patient may be ordered to drink Lemon Barley Water, and this drink is excellent for all invalids as it is both nourishing and thirst quenching. This recipe does not keep well, so make it fresh every day.

Ingredients: 1 oz pearl barley
$2\frac{1}{2}$ pints water
1 oz raisins
juice of 4 lemons

To mix: Wash the barley well and place in a pan with 2 pints water. Bring to the boil and cook for about 1 hour or until the liquid is reduced by half. Strain and add the raisins and the remaining $\frac{1}{2}$ pint water. Return to the boil and reduce the liquid to 1 pint. Strain and leave to cool. Mix with the lemon juice, chill and serve.

INDIGESTION

Remove the flesh from half a lemon and sprinkle it with iodised salt. Eat this before a meal to put the stomach in good order and to clean the mouth, tongue and teeth.

Another lemon remedy which claims to relieve even the worst attacks of indigestion instructs the sufferer to drink the juice of 1 lemon mixed with $\frac{1}{4}$ teaspoon bicarbonate of soda.

A glass of water with a little lemon juice is an excellent remedy for biliousness, and has been known to help pregnant women who are suffering from early morning sickness.

CURE FOR A HANGOVER

Mix the juice of a freshly-squeezed lemon with a spoonful of clear honey, and add 2 ice cubes. Drink, and repeat at hourly intervals.

HEADACHES

A cure for a headache that was advocated in the years before the First World War was to add the juice of half a lemon to a cup of sugarless black coffee.

SUN BLISTERS AND COLD SORES

Dab the affected area with fresh lemon juice as frequently as possible, using a clean piece of cotton wool each time.

LEMON BEAUTY

L emons carry with them an aura of freshness and cleanliness and, indeed, their acid juice has very real cleansing and bleaching properties. Our grandmothers certainly knew the value of the lemon as a beauty aid, and that knowledge is exploited by the cosmetics industry of today.

The citric acid in lemon juice helps to break down grease, and this is useful in making both shampoos and skin-care products. Its acidity helps achieve the correct Ph balance for particular skin types, and its bleaching action has been harnessed in preparations for blond hair. However, many of today's products tend to use stronger chemicals than those contained in the lemon.

The oil of lemon extracted from lemon skins has an attractive aroma which is used in beauty preparations and in perfumery.

But however attractive the commercial products sound, homemade beauty products cost a fraction of the price.

BLOND HAIR

When you wash your hair, add 1 teaspoon of freshly-squeezed lemon juice to the final bowl of rinsing water to help keep your hair really bright. If it's summer-time, this treatment (and the sunshine) will bring out the blond tints.

STAINED TEETH

Confirmed smokers can have real trouble with stained teeth, but this staining can be reduced considerably by rubbing the teeth with a paste made from lemon juice and 1 teaspoon bicarbonate of soda.

STAINED HANDS AND NAILS

Lemon juice will shift fruit and vegetable stains on the hands.

Remember to retain the lemon skins after squeezing them. Keep a couple on a saucer by the kitchen sink and dip your fingers into the remains of the flesh and rub over the stains every time you wash your hands. This treatment also works well for cleaning stained nails, and it strengthens them too. Use a cocktail stick dipped in lemon juice to clean under the nails.

FRECKLES

Use a mixture of equal quantities of lemon juice and water to tone down freckles and to whiten the skin generally.

SKIN SOFTENER

Mix together equal quantities of lemon juice and glycerine and rub into chapped hands or into patches of hard skin to soften them.

SKIN TONIC

Even if you do not have a really dull and greasy skin, your face will benefit from the aromatic oils in the lemon. This skin tonic will freshen your skin and help get rid of any blackheads or dirty pores. Do not use it more than once a month or your skin will get too dry.

Ingredients:	half a lemon boiling water a handful of herbs (rosemary, thyme, basil or mint)
To mix:	Pare off the rind from the half lemon and place in a large bowl. Fill up with boiling water. Add a handful of herbs and stir.
To use:	Cover your hair with a towel or a shower cap. Put your face over the bowl, about 12 inches away, and drape a towel over your head and the bowl. Keep your eyes closed and steam your face for about 15 minutes. Rinse afterwards with very cold water to close the pores.

SLIMMING WITH LEMONS

Lemons and lemon juice have a very low energy or calorie count, which makes them an ideal food and drink for slimmers. Undiluted lemon juice contains about 30 calories per 4 oz measure, and the quantities you are likely to eat or drink are considerably less than this.

CAUTIONARY TALE ON THE DANGER OF OVER-SLIMMING

There was a young lady of Lynn,
Who was so uncommonly thin
That when she essayed
To drink lemonade
She slipped through the straw and fell in.

LEMON BAKED FISH

4 white fish steaks (cod, halibut, haddock or hake)
1 in piece fresh root ginger, very finely chopped
1 shallot or small onion, very finely chopped
salt and pepper
1 lemon (½ sliced thinly, ½ squeezed)

Place each steak on a square of foil. Sprinkle with
ginger, onion and seasoning. Arrange the slices over
the fish and add a little juice.

Fold over the foil. Place on a baking tray and bake at
350° F/180° C/Gas 4 for 20–30 minutes.

Serves 4 *Calories per portion 250*

LEMON SALAD

Mix 1 teaspoon of grated orange and lemon rind with
the juice of 1 orange. Peel and slice 2 lemons, 1 orange,
3 tomatoes, and arrange them on a serving dish in
overlapping rounds. Peel and chop 1 avocado and
mix with lemon juice, then pile in the centre and
pour the orange juice and rind over the top.

Serves 4 *Calories per portion 30*

LEMON WATER ICE

This is a traditional coarse-grained Italian water ice known as Granita. It can be served in tall sherbet glasses or it can be spooned into frozen empty lemon halves.

1 pint water
4 oz sugar
1 tablespoon grated lemon rind
½ pint lemon juice (about 6-7 lemons)
sprigs of mint

Mix the water and sugar in a pan and heat gently to dissolve the sugar. Bring to the boil and simmer for 5 minutes. Leave to cool.

Stir in the grated lemon rind and the juice and place in a tray in the freezer compartment of the fridge. Leave to freeze solid.

To serve, shave off the mixture by scraping with a heavy spoon. Place in serving dishes or lemon halves and return to the freezer compartment until ready to serve. Decorate with sprigs of mint.

Serves 8–10   *Calories per serving about 65*

LEMON TEAS, CUPS AND COCKTAILS

The practice of serving slices of lemon in tea without milk originated in Russia, and is often referred to as Russian tea. A totally different way of serving lemon in tea is to infuse lemon peel in hot milk to flavour the tea, which is then served with sugar or syrup. This method is known as Bavaroise.

LEMON SHERRY CUP

The 17th century was a bibulous age in Britain and most people found straight lemonade rather insipid. So to pep it up a bit it was generally served with a little brandy and sack, a dry sherry-like wine from Spain. This modern version is ideal for a party.

2 pints lemonade (made as directed on page 40)
½ pint dry sherry
1 miniature bottle brandy
ice cubes (from one ice tray)

Garnish:
1 lemon, sliced
sprigs of mint

Mix the main ingredients together in a large bowl or jug. Serve in small wine glasses, and add a slice of lemon and a sprig of mint to each.

Serves 5-6 people (2 glasses each)

Lemon And Wine Sparkler

By the 18th century lemonade was tolerated as a non-alcoholic drink, but Englishmen still preferred to add an equal quantity of wine to their glass. Here is a party version that is excellent for a summer's evening.

2 pints lemonade (made as directed on page 40)
2 bottles sparkling white wine
ice cubes (from one ice tray)
2 lemons, sliced

Mix all the ingredients together and serve, strained, into champagne glasses.

Cocktail Time

Many cocktails use the zest of the lemon to bring out the subtle flavour of the mixture; others use lemon juice and sugar to frost the glass; yet more incorporate the juice in the mixture.

TIP FOR THE BEST ZEST

Pare off a very thin piece of rind with a potato peeler or a sharp knife, then carefully bend it back the opposite way to its natural curve. This helps to release the aromatic oils in the skin. Quickly pop it into the shaker and shake well, or add to the drink and serve.

LEMON COCKTAILS

This selection of popular cocktails uses lemon to counteract the sweetness of the liqueurs. Shake together with plenty of ice.

WHITE LADY
½ Gin
¼ Cointreau
¼ lemon juice

SIDE CAR
½ Brandy
¼ Cointreau
¼ lemon juice

SILENT THIRD
⅓ Whisky
⅓ Cointreau
⅓ lemon juice

WHISKY SOUR
¾ Whisky
¼ lemon juice
1 teaspoon beaten egg white
sugar to taste

SNAKE IN THE GRASS
¼ Gin
¼ Cointreau
¼ Dry Vermouth
¼ lemon juice

BLUE LADY
½ Blue Curacao
¼ Gin
¼ lemon juice

BALALAIKA
⅓ Vodka
⅓ Cointreau
⅓ lemon juice

MAPLE LEAF
⅔ Bourbon
⅓ lemon juice
1 teaspoon maple syrup

GREEN DRAGON
½ Gin
¼ Green Crême de Menthe
⅛ Kümmel
⅛ lemon juice

BRANDY GUMP
½ Brandy
½ lemon juice
2 dashes Grenadine

HOUSEHOLD
LEMONS

Make the most of the bleaching and cleansing properties of lemons around the house. On the bleaching front, lemons can be useful in removing stains and scorch marks from white linen – and this bleaching action is even more effective if it takes place in sunlight, as the heat and the sun's rays speed up the chemical reaction which takes place between the lemon juice and the stain.

Some commercial processes utilise the bleaching properties of lemon juice; calico printers find it brings out the white parts of patterns dyed with dyes containing iron.

The citric acid in lemon juice attacks grease, which is a reason why some washing-up liquids contain lemon juice, and it imparts an attractive smell, which manufacturers believe helps to sell the product.

Lemon also turns up in household polishes due to its clean fresh aroma.

STAINED WHITE LINEN

Rub a mixture of lemon juice and salt into the stain and leave to stand for a while; if possible, lay out the linen in the sun. Wash well and the stain should disappear.

To remove iodine stains, rub with freshly-cut lemons.

SCORCHED LINEN

If you are careless with the iron, light scorch marks on linen can sometimes be removed by rubbing with a cut lemon. Soak the material in water and wash again.

TO WHITEN LINEN

When boiling linen, drop two or three slices of lemon in the water.

NEW SHOES

Rub new shoes with a cut lemon to ensure first-class polishing thereafter.

COPPER AND BRASS

Household items made of copper and brass come up beautifully if rubbed with a mixture of lemon juice and salt. Rinse off the mixture with hot water and dry thoroughly with a clean soft cloth. Check first that the item has not been treated with a long-lasting polish or varnish as this treatment could remove it.

LEMON FOLKLORE
INDIAN LEGEND

According to an ancient Bengali legend, that country was the ancient home of all the ogres who used to terrorise the world. The story goes that they all lived in a single lemon fruit, and one day a little boy came along and picked the fruit and cut it up into a myriad little pieces. The ogres died and the world lived on in peace.

OLD WIVES' LORE

* Freshly cut slices of lemon placed on the window-sill will act as a barrier to flies.

* Dried lemon peel will make good firelighters; the oil in the peel will burn well.

* Lemon Treacle Posset will combat the common cold. Bring $\frac{1}{2}$ pint milk almost to boiling point, and add 1 tablespoon treacle and the juice of half a small lemon. Stir well and boil slowly until the milk curdles, then strain and serve very hot.

ORANGES AND LEMONS

Oranges and lemons,
Say the bells of St Clement's.

You owe me five farthings,
Say the bells of St Martins.

When will you pay me?
Say the bells of Old Bailey.

When I grow rich,
Say the bells of Shoreditch.

When will that be?
Say the bells of Stepney.

I'm sure I don't know,
Says the great bell at Bow.

Here comes a candle to light you to bed,
Here comes a chopper to chop off your head.

The rhyme 'Oranges and Lemons' is often sung as part of a game. Two players decide in secret which of them shall be an 'orange' and which a 'lemon'. They then form an arch and sing the song while the other players troop underneath. At the end of the song one of the players is captured and is asked (out of the remaining players' hearing) whether he will be an 'orange' or a 'lemon'. He then goes to stand behind the fruit of his choice, and the game continues. When the last player is captured there is a tug of war to see whether oranges or lemons are stronger.